The Life of Sir John Forbes (1787-1861).

Royal Physician, Medical Journalist and

Translator of Laënnec - a Victorian Polymath

by

Dr. Robin A. L. Agnew,

Emeritus Consultant Chest Physician, Liverpool.

The Life of Sir John Forbes (1787-1861).
First published by Bernard Durnford Publishing 2002
Second impression with corrections 2009
by Bernard Durnford Publishing
(*An imprint of Stewart Antill*)
This edition 2018 in a larger format
with additions and corrections
Bernard Durnford Publishing
(*An imprint of Stewart Antill*)

A catalogue record of this book
is available from the British Library

ISBN 978-1-904470-15-1

Format, design and cover by StewART

The Life of Sir John Forbes (1787-1861).

Royal Physician, Medical Journalist and

Translator of Laënnec - a Victorian Polymath

by

Dr. Robin A. L. Agnew,
Emeritus Consultant Chest Physician, Liverpool.

Bernard Durnford Publishing

The Life of Sir John Forbes (1787-1861).
Royal Physician, Medical Journalist and
Translator of Laënnec - a Victorian Polymath

by

Dr. Robin A. L. Agnew,
Emeritus Consultant Chest Physician, Liverpool.

*"Think where man's glory most begins and ends, and say
my glory was I had such friends".*

William Butler Yeats (1865-1939).

Dedication

To my Wife Ann: in appreciation for her
patience during the many hours that
I have spent on 'FORBES'.

Contents

Illustrations

Fig. 4 (p.45)   Sir John Forbes. Portrait by John
                Partridge.
                Now in the Royal College of Physicians,
                London, by kind permission of the College.

Fig. 5 (p.48)   Dr. James Johnson (1777-1845).
                Lithograph by T. Brigford, by kind
                permission of the Wellcome Library,
                London.

Fig. 6 (p.52)   Sir John Forbes in later life. Lithograph
                by T.H. Maguire, 1848, by kind
                permission of the Wellcome Library,
                London.

Fig.7 (p.57)   Memorial plaque of Sir John Forbes in
                St. Mary's Church, Whitchurch-
                on-Thames. By kind permission of the
                Rector, Rev. R. Hughes, with acknowl
                edgements to Sir Douglas Black, former
                PRCP Lond., and Mr. Robert Noble,
                Parish Archivists, Whitchurch and to the
                Editor, Journal of Medical Biography.

Preface

In early January 1953, a nervous young Irish doctor from Trinity College was interviewed, in the boardroom of the Royal West Sussex Hospital Chichester, for the post of house physician. Outdoors the weather was cold but the air was crisp and clear in the sunlight; fighter aircraft from the nearby RAF station at Tangmere droned overhead. The aspiring candidate, the present author, could not help but contrast the climate to the damp and foggy atmosphere of Dublin from which he had "travelled hopefully" overnight by sea ferry and rail.

He was invited to sit at the end of the long boardroom table for interview by members of the medical staff and lay officials; facing him at the other end of the table was the Chairman. However, what caught the candidate's eye even more than the intimidating gaze of the assembled company, was a portrait on the opposite wall: the subject's expression seemed to suggest a welcome to the formal proceedings of the Board. The framed picture showed the bespectacled features of a man in his late thirties with high cheekbones, clean shaven except for short side whiskers, wearing the dark jacket, winged collar and cravat of a typical Victorian doctor. Later, the author was told that the subject was "Sir John Forbes". Thus, in somewhat fraught circumstance, was my lifelong interest in a royal physician and medical journalist triggered in the year of the coronation of Queen Elizabeth II. But what was it about John Forbes that particularly struck my imagination?

It was explained to me in Chichester that their distinguished ex-physician had been the first to have translated the works of the immortal Parisian physician and stethoscopist, René Laënnec, from the original French, thus making them available to the English-speaking world. Later in his career, Forbes became an accomplished medical editor in London.

According to Sir William Wilde (1815-1876), in order to be successful as an editor of medical journals one must possess "untiring energy and perseverance, as well as the art of eliciting the knowledge and acquirements of others, together with a stern honesty of purpose, and suavity of manner". Did John - later Sir John - have these attributes? I will leave it to my readers of this narrative to judge for themselves.

On retirement from NHS duties, I was enabled to devote more time to the subject and to have had published some articles about the Scottish physician. In the course of writing these, I acquired some knowledge of John Forbes and his family background, which I thought might be used in publishing some form of literary tribute to him. This culminated in 1998, when I updated Parkes' valuable memoir of 1862 and had it printed for private circulation by the RSM Press. About the same time I was asked to bring up to date the entries for Sir James Clark and Sir John Forbes in the *New Dictionary of National Biography (New DNB)* to be published in 2004. Each was famous in his own right at a time when Medicine was emerging from the empirical therapies of 'bleeding and purging' to a more scientific methodology based on accurate clinical diagnosis and treatment. Some of the following narrative will appear under the title *'Sir John Forbes'* in the first edition of the *'New DNB'*; it is printed by permission of the Oxford University Press.

The lives of Forbes and Clark were intertwined from an early age as schoolfellows in their native Banffshire until the time that Clark finally visited Forbes just before Sir John's death in 1861. As fellow-Scots and as physicians they undoubtedly inspired each other throughout the trials and vicissitudes of their medical careers. On account of their life-long friendship, they named their only sons after each other: Clark naming his son John Forbes Clark and Forbes calling his Alexander Clark Forbes.

The sources of the material used for this monograph are listed in the select bibliography section together with Forbes' obituaries. Some of these tend to be lengthy and effusive in style but, fortunately for medical historians, they are valuable references! An example is my recent acquisition of a translation of the Scottish physician's notice of death that appeared in the *Medizinische Wochenschrift* of Vienna in November, 1861: this refers to him as the "modern Sydenham". [Thomas Sydenham (1624-1689), a distinguished London physician was known as the 'English Hippocrates']. The obituarist also mentions that Forbes was one of the founders of the "Sydenham Society", one of whose tasks was the publishing in translation of past and present medical works. The author of the obituary may have been the *Wochenschrift's* editor Dr. L. Wittelschöfer, but this remains speculative. For a description of the history of the original *Sydenham Society* (1843-1857) and the *New Sydenham Society* (1858-1907) and the roles played by Sir John Forbes and the young Jonathan Hutchinson (1828-1913), the reader is referred to *The Two Sydenham Societies* by G.G. Meynell (1985), Winterburn Books, Acrise, Kent p5.

The list of ships in which Forbes served during his career as a surgeon in the Royal Navy at the time of the Napoleonic wars is included as an addendum, as is the full text of Florence Nightingale's letter to Sir John of February, 1860.

The "Lady Flora Hastings scandal of 1839" refers to the unfortunate incident in the career of Sir James Clark when the naïve young Victoria asked for his opinion on a cause for the abdominal swelling of one of her ladies-in-waiting, whom the Queen suspected was pregnant. Sir James, mindful of the delicacy of the situation at court, decided to wait-and-see. Her ladyship died on the 5th July, 1839 of abdominal tuberculosis (Williams, 1949), although others have attributed her death to a liver tumour (Longford,1973). In a very full "Statement of the Case......," by Sir James Clark, written on 7th October, 1839, he pointed out that, in consultation with Sir Charles Clarke (an eminent contemporary obstetrician) during her ladyship's illness, they had very much doubted if she was, in fact, pregnant and they issued a certificate to her to that effect. Finally, the post-mortem examination had revealed extensive disease involving "...every organ within the abdomen...". My own opinion is that the exact diagnosis during life would have been very difficult, or impossible, to make without modern methods.

I have referred briefly in the text to John Forbes' visit to Ireland in 1852. He recommended the disestablishment of the Church of Ireland, showing that he was in tune with the general feeling that this was inevitable due to the decline in the numbers of the Anglican population. Disestablishment eventually took place in 1869, when William Gladstone (1809-1898) was Prime Minister, who was also keen to introduce Home Rule. At the same time, the Church of Ireland was disendowed and part of the proceeds used to progress the education of the Roman

Catholic majority. This rankled with the Protestants and it was said that there were some households, which would not allow Mr. Gladstone's name to be mentioned at the dinner table!

It is a pleasure to record the helpful advice of the late Mr. Bernard Durnford of Bramber, West Sussex, also my appreciation for the donation towards the cost of production received from the trustees of the Douglas Guthrie Trust of the Scottish Society for the History of Medicine. I should also like to add my tribute to Stewart Antill for all his hard work in producing this latest edition often at times of great personal hardship; without his dedication it would have been impossible to publish it. I should also like to thank Barry Hughes for his proof reading.

Acknowledgements

I should like to acknowledge the help that I have received from many sources and especially from personal communications. The late Dr. D. Evan Bedford (died 1978), Consultant Cardiologist to the Middlesex and National Heart Hospitals, wrote to me in 1967 with details of Forbes' "very interesting career". He suggested further research into aspects of Forbes' work in Chichester and kindly gave me the useful reference to Bishop's article in *Tubercle* (1961).

Mr. Michael Barrett has given me valuable archival material, including the Forbes' family tree. Figure 1. Alexander Clark Forbes (1824-1901) is buried at Whitchurch but his widow, who died in 1912 is buried at Chudleigh in Devon. Mr. Barrett has also acted as intermediary with the members of the Forbes family in Australia. Mr. David Forbes, FRCS, FRACS of Burradoo, New South Wales, has told me that he owns a signed copy of Sir John's book *'Sight-Seeing in Germany and the Tyrol in the autumn of 1855'*, which is dated June 20, 1856. The family also have a copy of *'A Physician's Holiday or a Month in Switzerland in the summer of 1848'* with "... written on the fly leaf in very legible ink is 'E.M. Forbes May 1849' ": this is the signature of John Forbes' wife, Eliza Mary. Also at Burradoo is a "most handsome tapestry fire-screen with Sir John's coat of arms and the motto *Labore Robore Spe*". I gratefully acknowledge this information and also for allowing me to see his eminent ancestor's private correspondence.

Mr. Robert Noble, Parish Archivist of Whitchurch-on-Thames, kindly arranged for the professional photograph of the memorial plaque in St. Mary's Church and also provided me with details of those in residence at Swanston House in the year of Sir John's death from the Electoral Return for 1861.

Mrs. M. Anderson-Smith of the Library of the University of Aberdeen has sent me important reprints of articles from the local literature on John Forbes and also her helpful comments on the handwriting in the margins of their copy of Parkes' 'Memoir' of 1862 relating to the early years in Dytac.

Mr. T. McCann has been equally obliging at the West Sussex Record Office and to him must be awarded the accolade of discovering the name of the artist of the Forbes' portrait in Chichester. Florence Nightingale's letter of 1860 to John Forbes is printed by permission of the West Sussex Record Office. I must acknowledge the aid of Dr. John Whiteside, the late Dr. John Mickerson and the late Mr. William Gammie, FRCS of St. Richard's Hospital in that City; Dr. Brian Owen-Smith of the Postgraduate Medical Centre there has also been most helpful to me. Miss Anne Blakeney has sent photographs of the "Forbes Window" and the present "Royal West" building.

Dr. Alex Sakula is a world authority on 'Forbes' and to him I am deeply indebted over many years for much information culled from his numerous erudite articles on medical history. The late Mr. P.J. Bishop, sometime Librarian of the Brompton (now the Royal Brompton) Hospital, was also an excellent source.

Mr. Geoffrey Davenport of the Royal College of Physicians, has been very helpful in allowing me access to the various Forbes' memorabilia in the Library of the College in London and also in viewing the family coat of arms, depicted in a stained glass window with the arms of other Fellows of the College, on the main stairway.

Mr. James Kyle, FRCS thoughtfully sent me the booklet *'A Pictorial History of Fordyce'*, by Christine Urquhart, from the *'Banffshire Journal'* on the *'George Smith Bounty'*. Dr. Robin Gatenby of the Aberchirder Medical Practice has sent me some good photographs of Fordyce village and school and Dr. Brendan Judge of Torquay has helped in trying to locate Forbes' house in Penzance and the family graves in Chudleigh.

My thanks are also due to Miss Alison Stevenson, Archivist of the Royal College of Surgeons of Edinburgh for the correct date of Forbes' gaining the Licence of the College on 18th February 1806 (not 1807, as stated in the standard biographies). Mr. Graham Salt has rendered invaluable assistance with research on aspects of Forbes' naval career and I should also like to acknowledge the help that I have received from the staff at the Public Record Office at Kew. Sir James Watt KBE, late Medical Director General (Naval) from 1972-1977, has been most generous with his encouragement over the years, particularly with naval medical history and useful sources of reference.

It is a pleasure to record the technical help of Richard Hancock and Tracy Bell of the Department of Medical Photography, Clinical Science Centre, Aintree Hospitals NHS Trust, Liverpool with preparing the illustrations for the booklet. *The Physician's Prayer,* part of which is quoted on page 54 is attributed to Sir Robert Hutchison (1871-1960), MD Edin.1896. It is printed from *Favourite Prayers*, compiled by Deborah Cassidi, London, Cassell (1998), by kind permission of Continuum Publishing Ltd.

The quotation on the title-page is from: *The Municipal Gallery revisited, New Poems*, 1938. In: Yeats's Poems, Edited and Annotated by A. Norman Jeffares with an Appendix by Warwick Gould, London (1989), PAPERMAC (a Division of Macmillan Publishers Limited), p 440.

I should like to acknowledge the valuable help with references given to me by Sarah Bakewell of the Wellcome Trust Library and also Frances Barendt. The latter's translation into English of the anonymous obituary notice in the Viennese *Medical Wochenschrift* drew my attention to some aspects of Sir John's career of which I had previously been unaware. I am also grateful to Professor Karl Holubar, MD, FRCP of the Institute for the History of Medicine in the University of Vienna for his advice on the identity of the author of the notice. I should also like to acknowledge the financial contribution to the publication of this book by Dr & Mrs A Barendt of Helsby, Cheshire. Finally, to all those whose names are not mentioned, I apologize: your contributions are equally appreciated.

FORBES FAMILY TREE

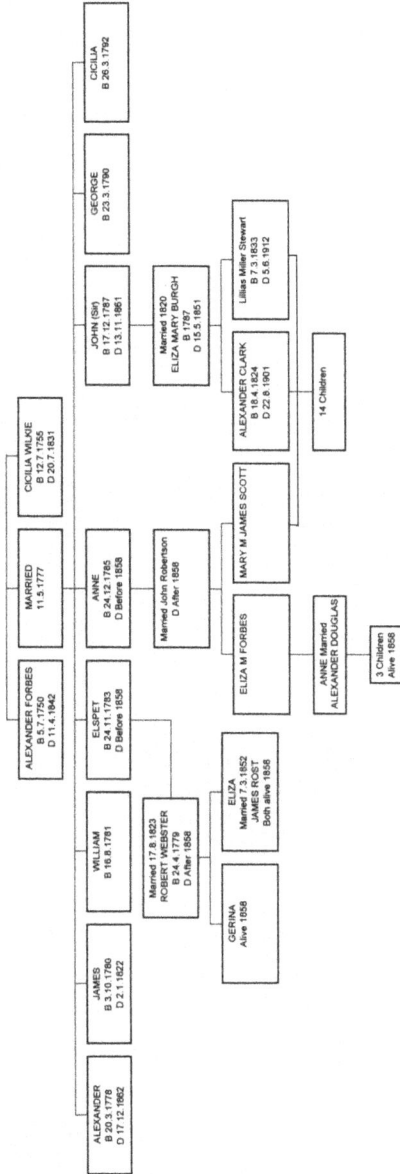

Fig.1

21

Chapter 1
Early Years and Education

John Forbes was born on 17 December 1787 at Cuttlebrae, near Cullen, in the parish of Rathven, Banffshire. He was the fourth son of Alexander Forbes (1750-1842), a local tenant farmer and Cicilia [sic] Wilkie (1755-1831). Two years later the family moved to nearby Dytac in the parish of Fordyce, where John received his early education at the Parish School from the age of 6 to 15. Alexander Forbes farmed land on the Findlater Estate near Kilnhillock, which became the home of eleven-year-old James Clark (1788-1870) in 1799. Over the next three years the young John Forbes and James Clark walked together to Fordyce School, later to become Fordyce Academy. They learned English, Modern Languages, Greek and Latin Grammar; for recreation they swam in the sea at Sandend on the coast, at which activity young Forbes excelled. It is not surprising that Forbes and Clark became lifelong friends. In later life their medical careers overlapped; and it was James Clark who encouraged Forbes to translate Laënnec's '*De L'Auscultation Médiate*' into English in 1821.

At the age of 15, having obtained a bursary founded by an ancestor of his mother at Fordyce, John attended Aberdeen Grammar School, spending a year in the Rector's class in order to extend his knowledge of English, French and the Classics. His time there did not coincide with that of George Gordon Byron (1788-1824), who was also a pupil at the Grammar School in Aberdeen from 1794 until he left on inheriting the title of Lord Byron in May, 1798. Curiously, the young Byron was taken by his mother on holidays to Banff where, in spite of his clubbed foot deformity, he may have learned to swim in the sea. Banff - "a pleasant town on the north-east coast" - is situated only a few miles along the coast from Sandend Bay, where young Forbes acquired his aquatic expertise.

From Grammar School John Forbes entered the Arts course at Marischal College of the University of Aberdeen, where he attended classes for two years between 1803 and 1805 but there is no record that he ever graduated. To travel, he walked from his home in Banffshire to Aberdeen. In what little spare time there was available, he developed a medical interest by becoming apprenticed to two medical practitioners in Banff. The war with Napoleon was then raging and, possibly inspired by Nelson's victory at Trafalgar in October 1805, Forbes decided to enlist in the Navy. He proceeded to tuition in Edinburgh and obtained the Diploma of the Royal College of Surgeons there in February 1806. In the following year, owing to the shortage of naval surgeons at that time, he was able to enter the medical service of the Royal Navy in the rank of Temporary Assistant Surgeon. In order to join the fleet, the 20-year-old Scot sailed from Leith in a fishing smack, the voyage to London taking 14 days; thence, travelling by coach for a further 3 days and nights, he joined the 100 gun three-decker HMS ROYAL GEORGE at Plymouth.

Chapter 2
Naval Career (1807-1816)

He served in the Channel Fleet for two years and was confirmed Full Surgeon on 27 January 1809. He then joined the smaller 12 pounder, 32 gun frigate HMS CASTOR to serve on the Leeward Islands station in the West Indies. While on passage to Admiral Cochrane's squadron in the Caribbean his ship was involved, together with HMS POMPÉE, in a long running sea engagement culminating in the capture of the French 74 gun line of battle ship D'HAUTPOULT after a brief but bloody action on 17 April 1809. On account of his knowledge of the French language, Forbes was ordered by the Admiral to transfer to the stricken vessel to assist with caring for the wounded. Following this he served in various ships including the 18-gun brig HMS NETLEY and was present at the capture of Guadeloupe by the British in February 1810. Following shore leave at home he joined the staff at the Royal Naval Hospital Haslar where he was Assistant Medical Officer from 6 January to 7 April 1811. During his spare time at sea, Forbes had improved his basic knowledge of French and other European languages. He had been a strong swimmer since his schooldays and had saved the life of a shipmate in the Caribbean and, later, of a rating washed overboard into the Elbe from HMS DESIRÉE, 36 guns, serving on the North Sea station in 1813. On this occasion Forbes dived in and nearly lost his own life as he was swept for two miles down river by the strong current before being rescued. A further example of his courage took place when he volunteered for a 'cutting-out operation' but the officer in charge decided not to risk the lives of his crew by rowing within range of enemy fire; Forbes was furious and, throwing down his sword in the bottom of the boat, declared vehemently that "... he would not be caught on a fool's errand again".

Forbes has been described at this stage of his career as being "...about the middle height, and was strongly and squarely built; he had blue eyes, a light florid complexion and was full of spirits, frank and joyous. His manner was bluff and hearty, but pleasing from the evidence it gave of sincerity and goodness. His habits were extremely active". Evidence of a more sensitive side to his character may be deduced by his being said to have written at sea some amateur poems, which were published in the local West Indian papers but there is no record of these.

Forbes served for two years (1811-1813) on the North Sea station at a time when Emperor Napoleon Bonaparte (1769-1821) had suffered a major defeat in Russia. At the same time in the Iberian Peninsula, the Duke of Wellington (1769-1852) was showing that, in the words of George Bernard Shaw (1856-1950) "An English Army led by an Irish general..." was more that a match for the French. In 1813, when Forbes' warship was in the estuary of the River Elbe, a curious incident took place which nearly cost him his life. As he was the only one on board who could speak French, he was ordered ashore to go to Bremen with despatches for a Russian general in command of 10,000 Cossack troops. The Scottish doctor was well aware of their fierce reputation but they were very friendly towards him, although they had a strong hatred of the French. His escort of ten Cossacks on the journey surprised two French gendarmes and proceeded to cut them to pieces. When Forbes tried to intervene to save them, one of the Cossacks threatened him with his lance, so that he had to stay his hand. Later, however, they became very congenial again towards him, as if nothing had happened. They expressed a strong wish to force march across France in order to join the 'Iron Duke' at the Pyrenees, declaring Wellington to be the finest general in Europe!

The Scottish surgeon combined very well the roles of medical and naval officer and was particularly methodical and reliable. These qualities were rewarded by his appointment in April 1814 as flag surgeon and secretary to Rear Admiral Philip Charles (later in 1815, Sir Philip) Durham. Durham flew his flag in HMS VENERABLE, 74 guns, as Commander-in-Chief of the Leeward Islands station from 1814 to 1816. Forbes was present and played an important part as Admiral's secretary in the capture of Guadeloupe from insurgent French Bonapartists in August, 1815. This was a hazardous amphibious operation against a determined foe, carried out against a background of the impending hurricane season. It deserves the epithet contained in Virgil's Aeneid: *Audentis Fortuna iuvat* - Fortune favours the Brave. (See select bibliography). And page 133 in *LIGHTFOOT WINDS* ISBN 978-1904470052 also by Robin Agnew

Forbes left VENERABLE on 30 May, 1816, when the ship 'paid off' on arrival home at Portsmouth. This was his last appointment in the Royal Navy.

He would have been entitled to an award of the Naval General Service Medal (NGS) from being present at several naval engagements but there is no family record. [Personal communication, David Forbes, NSW, Australia].

Chapter 3
Edinburgh and Penzance
(1817-1822)

In 1816, at the end of the Napoleonic Wars, Forbes was put on half pay. He returned to Edinburgh as a 29-year-old mature student and studied at the medical school and University. In August 1817 he received his Doctorate of Medicine on the same day as James Clark. The title of his inaugural dissertation 'Tentamen Inaugurale de Mentis Exercitatione et Felicitate exinde Derivanda' reflects his Scots perception of life: his Latin thesis was dedicated to Philip Charles Durham (1763-1845) in the flattering terms in common use at the time but its main theme - on the satisfaction to be gained from mental pursuits - was the keynote of Forbes' philosophy.

As an example of his wide interests, he had attended lectures in geology by Professor Robert Jameson (1773-1853), while pursuing his medical studies. The professor was asked to recommend an Edinburgh physician with an interest in geology for a vacancy in medical practice in Penzance, resulting from the retirement of Dr. J.A. Paris (1785-1856). John Forbes MD was duly appointed and moved to Cornwall in September, 1817. There was also some speculation that his move was influenced by health reasons, as he was known to have suffered from chest symptoms in later life. Pulmonary tuberculosis was not uncommon in the Navy: it is possible that the Scottish physician may have contracted this infection during his naval service but overcame it by his strong constitution.

Dr. John Forbes, already a naval surgeon, now used his medical expertise acquired at Edinburgh as a physician to Penzance Public Dispensary from 1817 to 1822. He also worked in general medical practice and saw patients with a wide variety of diseases in all age groups throughout the county of Cornwall, including the Isles of Scilly. In particular, he described the stethoscopic signs of pulmonary tuberculosis in two Cornish underground miners, a group in which TB was then not uncommon. He would not have been taught by his professors at Edinburgh about stethoscopy as the monaural instrument had only been invented in 1816 by René Laënnec (1781-1826). [For the description of Forbes' notes on patients seen by him at Penzance, I am grateful to the late Dr. John Craig (1899-1977), formerly Professor of Child Health in the University of Aberdeen].

During his five years at Penzance, Forbes laid the foundations of his knowledge of the stethoscope. His friend James Clark had given him the instrument following Clark's visit to Paris in 1818. To quote a footnote of Clark: "An English [sic] physician to whom I gave a sthenoscope [sic] which I brought from Paris with me, informs me he has already found it useful in the diagnosis of some of the diseases of the heart". Clark was enthusiastic about "Dr. Laënnec's valuable work", entitled *'De L'Auscultation Médiate...'* (1819) and was the instigator of Forbes' classical translation into English: this 1821 edition was dedicated to Matthew Baillie (1761-1823). Forbes has been criticised for rearranging and shortening the original text and especially for altering the terminology of Laënnec's description of the lung sounds found at auscultation.

28

The first edition, written with a quill pen, from the Chapel Street Dispensary in Penzance, was a great success and inspired three further editions in 1827, 1829 and 1834, published after Forbes moved to Chichester in 1822. These editions were all dedicated to James Clark. Four years prior to the second edition, Forbes had written to Laënnec in Paris apologising for the "great liberties" he had taken in his translation or, as he admits "...my ABRIDGEMENT (I will not say TRANSLATION) of your immortal work". Forbes' explanation was that a more concise version would make a greater impact on British readers. The second edition (1827) contains some biographical notes on Laënnec, who had died in 1826. (Fig 1A). A third edition appeared in 1829, very similar to the second, and a fourth (1834) based on Mériadec Laënnec's third French edition (1831) of his late cousin René's work. It was in this 1834 edition that Forbes changed the translation of Laënnec's 'râle' from the English 'rattle' to the Latin 'rhoncus'.

These translations, however imperfect they appeared at the time, were paramount in spreading the teachings of Laënnec on diseases of the heart and lungs by the use of the stethoscope to the English-speaking medical world. Five hundred copies of Forbes' 'A Treatise on the Diseases of the Chest' (1821) were sold by the end of 1823.

As well as working as a physician in Penzance, Dr. Forbes took a keen interest in local affairs and helped found the Public Library in 1818. He also developed his interest in geology, acting as secretary of the Royal Geological Society of Cornwall. He contributed two papers: *'On the Geology of Land's End district'* and *On the Geology of St. Michael's Mount'*, which were published in the *'Transactions of the Royal Geological Society of Cornwall'* in 1822. *'Observations on the Climate of Penzance and the District of*

A TREATISE

ON THE

DISEASES OF THE CHEST,

&c. &c.

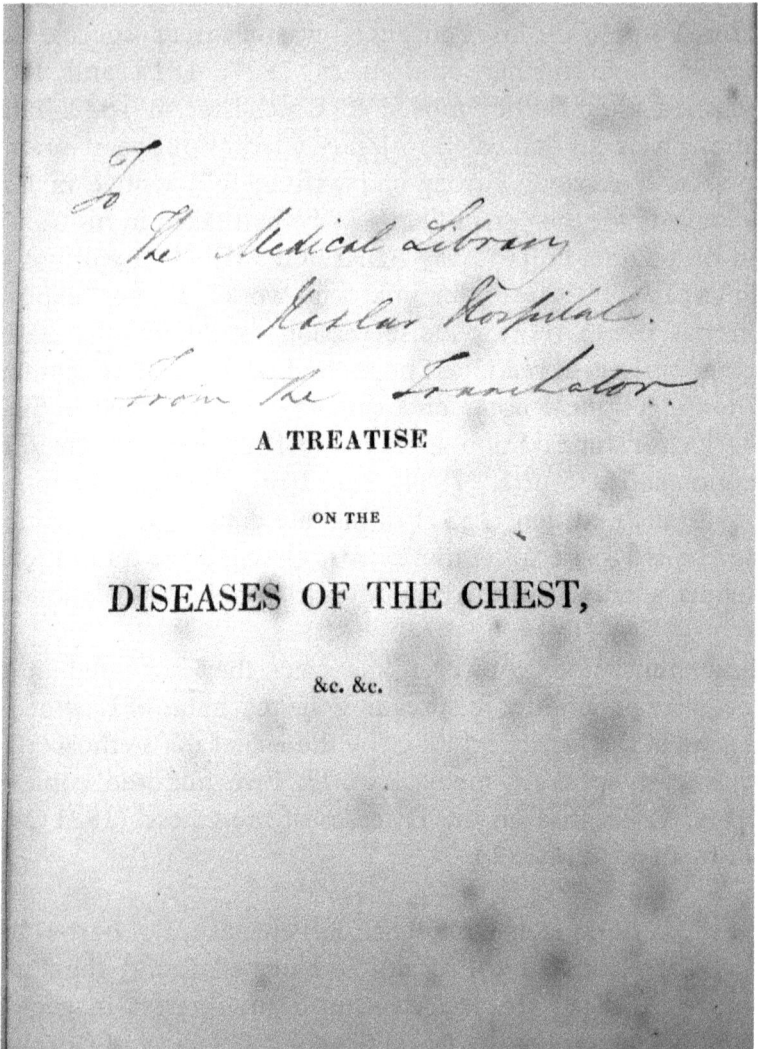

Fig.1A: Image from the Historic Collections Library
of the Institute of Naval Medicine by kind permission of the Officer in Charge.

Land's End in Cornwall; with an Appendix containing Meteorological Tables, and a Catalogue of the Rarer Indigenous Plants' was read before the Penwith Agricultural Society and published by the request of the members at Penzance in 1821. This testified to his skills as a Naturalist and, also, as an observer of local climatic conditions, to his naval experience. In the preface, he hopes that the publication will stimulate other medical practitioners to similar studies and to recommend Penzance for convalescence to their patients. A further paper *'On the Temperature of Mines'* was printed in the *'Transactions'* in 1822. This concerned the health of Cornish tin and copper miners, especially their working conditions. It was later said by a Professor of Hygiene in London to have been one of the best papers ever written by Forbes.

While at Penzance he married Eliza Mary, the daughter of the late John Burgh, a partner in a bank in Calcutta, who had died in 1793. Eliza was born there in 1787 and with her widowed mother, née Elizabeth Mary Cumberledge, had returned to live at Great Torrington in Devon. The Burghs, or De Burghs, are descended from a Norman-Irish family from Dromkeen, County Limerick. The marriage between Eliza Mary Burgh and John Forbes took place at the parish church of Great Torrington, on the 19th May, 1820, "by licence". The bride was the same age as her husband; one son was born in Chichester on 18th April 1824, Alexander Clark Forbes; he was named after his paternal grandfather, his uncle Alexander, and his father's good friend, James Clark.

Chapter 4
Chichester (1822-1840):

In 1822, John Forbes and his bride of two years moved from Penzance to Chichester to succeed Dr. William Burnett (1779-1861). Burnett had been urgently recalled to Haslar Hospital to improve medical services and later became Physician General to the Navy. The Forbes' lived at No. 21 North Street not far from where the poet John Keats (1795-1821), staying briefly with friends at the far end of East Street, had started to write his poem *The Eve of St. Agnes* in January, 1819. As at Penzance, Dr. John Forbes soon became involved with non-medical coteries. In 1831, he founded the Chichester Literary and Philosophical Society to which he later contributed several papers; he also published two articles in the *'Transactions of the Provincial Medical & Surgical Association'*. The first appeared in 1834, entitled a *'Sketch of the Medical Topography of the Hundred of Penwith, comprising the district of the Land's End in Cornwall'*, followed two years later by a further continuation, which included some pioneer studies on the diseases of underground miners.

Forbes built up a large practice in the cathedral city, extending over a broad area of West Sussex and East Hampshire and earning him as much as £1,500 a year (about £70,000 at today's values).

A potential rival physician was John Conolly (1794-1866), also an Edinburgh graduate (MD 1821). In the event, they formed a strong friendship and, wisely, Conolly departed amicably to Stratford-on Avon, where he remained until 1827. They had met socially at the home of the doyen of the Chichester medical fraternity, Dr. T. Sanden. He had been one of the founders of the Chichester Public Dispensary in 1784 but, by the time of Forbes' arrival in 1822, the building was out of date. It was largely through John Forbes' fund-raising efforts that it was replaced by a modern infirmary block, which opened for patients in 1826 at a cost for building of £4,000 approx. This became the Royal West Sussex Hospital in 1913.

Forbes became popular both as a man and as a physician in hospital and domiciliary work. As proof of the latter, there exists a silver communion cup, with an inscription: *'Sick and you/Visited me/to/Doctor Forbes/from* J.P.' the words are surrounded by an inscribed wreath. It was later donated by Forbes to the hospital. The cup, late 18th century silver, was in regular use there as recently as 1960.

Within two years of his arrival at the old Dispensary, Forbes had written an important work entitled *'Original Cases with Dissections and Observations illustrating the use of the Stethoscope and Percussion in the Diagnosis of Diseases Of The Chest; also commentaries on the same subjects selected and translated from AVEN-BRUGGER, CORVISART, LAENNEC and others'.*
This was published in London in 1824 and dedicated to RTH LAËNNEC in Latin. The book contains Forbes' further experience in the use of the stethoscope and also includes the first translation into English of the 'Inventum Novum' of Auenbrugger (1722-1809). Besides auscultation, Forbes used the technique of percussion of the chest wall based on the

Austrian doctor's original work published in 1761, which had been translated into French by the Parisian physician Corvisart (1755-1821). There are descriptions of the case histories and physical signs of 39 patients seen at Chichester by Forbes personally; these signs are verified by autopsy in those whose disease proved fatal.

The principle of trying to link the vital signs with post-mortem findings was new in Britain and further extended the teachings of Laënnec. *The Lancet,* in a favourable review, commented that "... it is the first of the kind published in the country, and reflects no small credit on the author for the pains he has taken to make these discoveries generally known". The publication was also warmly received in Edinburgh. There was a need, at the time, for a reliable textbook on the practice of auscultation and the dissemination of Laënnec's views to a medical world still sceptical about the value of a stethoscope in clinical diagnosis-a view initially held by Forbes himself. A fifth edition of his translation of Laënnec's epic work on auscultation was published in New York in 1838.

Chapter 5
Medical Journalism

John Conolly had left Chichester in 1823. He had kept in touch with Forbes and another Edinburgh graduate, Alexander Tweedie (1794-1884) who practised as an ophthalmic surgeon in London, where Conolly had briefly held a Professorship of Medicine at University College. At Tweedie's suggestion, the three combined to produce the *'Cyclopaedia of Practical Medicine'* but the main Editor was John Forbes. This appeared in 4 octavo volumes between 1832 and 1835. It was published monthly in five-shilling numbers in London, Edinburgh and Dublin and also in the United States of America. It proved a popular forum for the best medical writers in the British Isles and was eventually sold off in 1835 at a handsome profit for the publishers, Sherwood, Gilbert & Piper.

Forbes had written articles for the *'Cyclopaedia'*, including an essay on 'Auscultation', but a more valuable contribution was his compilation of an excellent bibliography of its contents, the first of its kind: *'A Manual of Select Medical Bibliography'*, London, 1835, remedied the absence of references from the individual articles and his arrangement of them in alphabetical and chronological order form a yardstick of excellence in medical literature.

Forbes was elected a Fellow of the Royal Society in 1829 and, perhaps encouraged by this signal honour, applied for the post of Professor of the Nature and Treatment of Diseases at London University in 1830. Among the impressive list of testimonials, contained in Forbes' personal papers, submitted to the Council in support of his application is one from 'the late Matthew Baillie, M.D. Physician to the King', dated 'Feb. 16, 1822'. Baillie (1761-1823) had been the author of the first treatise on morbid anatomy (1793); Forbes had wisely obtained this testimonial in the year prior to Baillie's death. Others were from his former professors at Edinburgh, Home and Duncan, as well as from Sir James McGrigor (1771-1858), Director General of the Army Medical Department and, not surprisingly, one from James Clark M.D. 'Physician to St. George's Infirmary, London'. In spite of these, all laudatory appraisals, and a more cautious commendation from the President of the Royal Society, Davies Gilbert, his application was unsuccessful.

In 1836, Forbes and Conolly embarked on a new venture in medical journalism. This was entitled the *'British and Foreign Medical Review, or Quarterly Journal of Practical Medicine and Surgery'*. In April, 1839, Conolly resigned in order to return to full-time work in psychiatry at the Middlesex County Asylum at Hanwell. They remained in close touch and were conferred with the Fellowship of the Royal College of Physicians on the same day, 6 July 1844. John Conolly had been associated with Forbes in helping "their mutual friend", Dr. Charles Hastings (1784-1866) in founding the Provincial Medical and Surgical Association in 1832; this was the forerunner of the British Medical Association. Charles Hastings (MD Edin., 1818) was knighted in 1850.

The *'Review'* (1836-1847) was an even more prestigious journal than the *'Cyclopaedia'* (1832-1835). The first number appeared in January, 1836 and, after 1837, was published by John Churchill, with whom Forbes established a good working relationship to their mutual advantage. A rival publication was the '*Medico-Chirurgical Review*' (1816-1847), whose editor was James Johnson (1777-1845). The '*Medical Review*' was a great success, both in Britain, Continental Europe and America. Most of the editorial work was done by Forbes, even before the departure of Conolly. Parkes was convinced that Forbes' '*Review*' did more than any other journal to enhance the reputation of British medicine both at home and abroad. It became accepted and read all over Europe and America as the articles selected by the editors helped to promote more rational methods of treatment than the bleeding and purging still prevalent. Forbes was sole editor for eight years and retained the identity of his anonymous reviewers during his lifetime.

Parkes has described Forbes' benevolent character and philanthropy during the prime of his life at Chichester. (He was a patron of the Cicestrian 'shoemaker poet', Charles Croker). At that time he was described as being: "... active, lively, full of energy, remarkable for the elasticity of his step, and his almost youthful bloom and freshness of complexion...". This picture fits in well with his portrait at that time: he was persuaded by his friends in Chichester to sit for a local artist of sound reputation, James Andrews, who has succeeded in depicting his intelligent bespectacled features precisely. The portrait, previously in the boardroom of the Royal West Sussex Hospital, now hangs in the foyer at St. Richard's Hospital, Chichester.

Forbes had been thwarted in his bid to achieve academic status at London University in 1829 but, two years later was appointed Physician-in-Ordinary to the Duke of Cambridge.

Fig 2: John Forbes MD as a Physician at Chichester.

From 1836 he was deeply involved with the editorship of the *'Review'* and finally made the decision to exchange the tranquil surroundings of a cathedral city and his profitable position there for the hazards of medical journalism and private practice in the Capital. His resignation as senior physician at the Infirmary was announced on 15 October 1840. The Governors expressed their appreciation of his "skill and humanity" as a physician and of his "zeal and assiduity" in placing the Infirmary "... in its present state of respectability and usefulness". Forbes moved from Chichester to London in what was to prove a turning-point in his career. It was also an anxious time for his wife and 18-year-old son, Alexander. The Family took up residence at 12 Old Burlington Street, between Old Bond Street and Savile Row. At this critical stage, they were greatly helped by John's old friend and schoolmate from Fordyce, James (now Sir James) Clark, whose medical career had barely survived the Lady Flora Hastings scandal in 1839. A vacancy occurred as Court Physician to Queen Victoria's Household; almost certainly due to Clark's influence with the Queen, Dr. John Forbes was appointed: this honour, dated 15 February 1841, was timely. Forbes now concentrated on the sole editorship of the *'Review'*, which soon established his high reputation in the world of medical journalism. He also built up his consultant practice, enhanced by undertaking medical examinations for life assurance companies including the Irishman Daniel O'Connell (1775-1847) MP, a skilled barrister and politician, who advocated non-violent means to achieve his aims. He had now reached the peak of his medical and journalistic careers in London. Unfortunately, the success of these years was marred by the chronic ill-health of his wife, Eliza, who died in 1851, aged 64, and was buried at Kensal Green Cemetery. Charles Croker, who had written a eulogistic sonnet to Forbes, 'a friend though ten long years', on his leaving for London in 1840, later contributed six stanzas in memory of Eliza Mary which were published in 1860.

There is also a lovely stained glass window in Chichester cathedral in memory of Sir John and his wife, who founded the local Dorcas Charity.

Fig 3: Stained Glass Window at Chichester Cathedral.
In memory of Sir John Forbes and his wife

John O'Neill (1778-1858) was an impoverished Irish shoe-maker/poet who visited John Forbes at No.12, Old Burlington Street in the early 1850's. The doctor was a member of the Royal Literary Fund (RLF) and a fervent supporter of the temperance movement. Due to his poor financial circumstances O'Neill had applied for relief to the RLF, of which Forbes served on he committee. (The RLF members were – and still are – literary gentlefolk with a benevolent fund whose purpose is to help British authors in financial straits. Founded in 1790 by the Reverend David Williams, it was inspired by the death in a debtor's prison of a translator of the works of the early Greek Philosopher Plato, Floyer Sydenham (1710-1787). Clearly, as a fellow translator, the aims of the RLF stirred the philanthropic aspects in Forbes' character just before he embarked on his own investigative journalistic tour in Ireland in the Autumn of 1852)

John O'Neill was the author of a temperance poem entitled The Drunkard (1840; second edition 1842). Forbes made a very favourable impression on the Irishman, who remarked on the 'the distinguished scholar and gentleman' as assuming "no superiority in his manner and bearing over the poor uneducated cobbler. I left Old Burlington Street both gratified and grateful, believing that there was more humanity and generous sympathy for the poor among the higher classes of society than the world gave credit for". Before leaving, the Scottish physician presented O'Neill with a copy of Pain's Task of the Age and asked him to read it and to report back and offer his opinion on the work. He also made a generous offer to buy twenty copies of The Drunkard at one shilling each, for which he "presented me with a sovereign, telling me I need bring only four copies for himself and his friends". They parted with expressions of mutual friendship and respect and a request by Forbes "any morning you are coming this way, I would be glad to see you".

In addition to his other work, the Scottish physician continued to write for his new publisher, John Churchill: *'Illustrations of Modern Mesmerism, from Personal Investigation'* appeared in 1845. The author is proudly described as *'Physician to Her Majesty's Household'*, with an appropriate quotation from Shakespeare 'All My reports go with the modest truth...' (King Lear, iv,7). In the

41

preface Forbes confesses to being dubious regarding "Clairvoyance and other Mesmeric wonders" but felt it his duty to record some details about Mesmerists and their performances, at that time in fashion. His notes were based on contributions to the *'Atheneum'* [sic] and the *'London Medical Gazette'* and included some humorous and sceptical examples of demonstrations. Two other publications by Forbes were *'Mesmerism True-Mesmerism False'* and an article *'On Mesmerism'* in the April, 1845, number of the *'British & Foreign Medical Review'*; this summarized the literature and recorded Forbes' failure "to obtain any advantage from mesmerism in several instances where we have employed it...". It was a topical subject in early Victorian medicine, which was, by no means, generally accepted but Forbes felt that its therapeutic effects should be the subject of a scientific trial.

He had published an article *'On Sleepwalking, Clairvoyance and Animal Magnetism'* written in German, in collaboration with A. von Hummel, which appeared in Vienna in 1846 and was also interested in establishing the truth or error of such 'fringe' subjects as Phrenology and Homeopathy. John Forbes was awarded an honorary Fellowship of the Imperial Society of Physicians of Vienna on 26th March, 1845.

Chapter 6
Forbes and Homeopathy

In January, 1846, there appeared in the *'British & Foreign Medical Review'* an unsigned commentary referring to a review of nine articles on Homeopathy by various British and Continental authors. Its title was *'Homeopathy, Allopathy and "Young Physic"'*. In twenty sections (pp 262-65) the author, almost certainly the Editor, sets out the case for the "vis medicatrix naturae" and the shunning of polypharmacy, especially by inexperienced doctors; there is also a plea for medical students to think for themselves and not to be afraid of questioning dogmatic teaching. Forbes concluded that the practice of Medicine must be based on a combination of art and science and that this could only take place by substantially improving the standard of teaching of all doctors. Such opinions were considered too iconoclastic by the medical establishment of the day, who claimed that Forbes' views favoured the system of Samuel Hahnemann (1755-1843) based on the belief of 'similia similibus curantur', although this was not the author's intention. He appears to have kept an open mind on the principle of 'like cures like' and certainly had no time for medical humbug.

The sour and bigoted interpretation of his article was reflected in an unflattering obituary notice of Forbes, which appeared in the *'Lancet'* nearly sixteen years later. His reputation as an unbiased editor and his integrity as a physician were not harmed, as in the same year 1846, the author of the "obnoxious articles" (to quote the *'Lancet'*) in his *'Review'* was appointed one of the first two consulting physicians to the 'Brompton Hospital for Consumption and Diseases of the Chest'.

News of the use of ether anaesthesia for a molar tooth extraction arrived in England on 16th December 1846, when the transatlantic mail ship Acadia docked in Liverpool. [Personal communication, Dr Anne Florence, Liverpool] RMS Acadia was one of three sister-ships built by Cunard for a fortnightly mail service to Boston.

The severe criticism of his controversial article, which was much more than a review, may have lead to Forbes' resignation as Editor in the 61st year of his life. Before leaving in 1847, however, he had the opportunity of a final journalistic scoop. Following correspondence with medical friends in Boston, Mass. dated November 1846, in which the benefits of ether anaesthesia were attributed to Drs. Jackson and Dental Surgeon Morton, Forbes saw for himself this new technique at University College Hospital in London on 21st December. The intrepid Scottish surgeon, Robert Liston (1794-1847), performed an amputation of the thigh of a man, who had been anaesthetised by "sulphuric ether vapour". This made the operation quite painless and the patient quickly recovered consciousness. Forbes hurried to his editorial office to record, in time for the New Year number of the *'Review'*, this historic first use of a general anaesthetic in England: he was convinced that this *'New Means of rendering Surgical Operations Painless'* would take over the role of the "mesmerists" in the control of feeling pain.

Forbes was a confirmed teetotaller: his article on 'Temperance and Teetotalism: An inquiry into the Effects of Alcoholic

Drinks on the Human System in Health and Disease' was published in London in 1847 as a sixpenny pamphlet by John Churchill. It was reprinted from volume 48 of the 'British & Foreign Medical Review' just before the editorship changed hands. The new editor, Dr. W.B. Carpenter (1813-1885) was also a leading light in the temperance movement and an eminent physiologist; he served as editor until 1852.

Forbes' *'Review'* had initially been self-supporting but later incurred a loss amounting to about £500 (£25,000 today). Even so, at the time of his death some fifteen years later, informed opinion was that he had succeeded in bringing about the reforms so vigorously and honestly enunciated in his article on 'Young Physic'. In a valediction in the last number of the *'Review'* edited by him, Forbes set down the principles which had inspired him over the twelve years of his ownership. In particular, he stated that his determination had been always to preserve an impartial approach to his editorial duties. He mentions the administrative and financial problems encountered and gives a spirited and logical justification for his polemical article on *"Homeopathy..."*, which may have contributed to the journal's commercial failure.

In a final postscript Forbes announced the amalgamation of the two London Quarterly Journals into the *'British and Foreign Medico-Chirurgical Review'*. He paid tribute to his publisher, John Churchill, who was now one of the owners of the new Journal, which would continue "... the same independent and liberal course..." as its predecessor. As a token of their affection on his leaving the 'Review', he was presented with a splendid candelabra in the names of 264 physicians and surgeons of Britain and America, many of whom had been former contributors and remained his friends, as well as his portrait painted by John Partridge (now in the Royal College of Physicians, London)

Fig 4: Sir John Forbes.

The *'Medico-Chirurgical Review'*, with which Forbes' journal combined in 1848, was edited by Dr. James Johnson (1777-1845). Ten years younger than Forbes, he was born on a small farm in the North of Ireland. After a rudimentary education, he moved to London where he supported himself as an apothecary's assistant and passed the examinations of the Surgeon's Hall in 1798. He then enlisted as surgeon's mate in the Royal Navy and was promoted to full surgeon in 1800. Following wide experience at sea, including the Far East, he wrote a book on the influence of tropical climates on the health of seamen, first published in 1812. Having retired from the Navy, he graduated MD at St. Andrews and was admitted a Licentiate of the College of Physicians in 1821. He settled in London and became a very successful editor, his journal averaging 2,500 copies at the height of its circulation. However, its popularity declined and, in later years was overtaken by Forbes' *'British and*

Foreign Medical Review, or Quarterly Journal of Practical Medicine and Surgery' . Earlier, Johnson had been physician extraordinary to William IV (1765-1837), the 'sailor king'. James Johnson died in 1845.

Fig 5: Dr. James Johnson.

Chapter 7
Further Career

The Forbes family continued to live at 12 Old Burlington Street after his retirement in 1848. Dr. Forbes was a regular supporter of the theatre, especially Shakespearean drama; he was fond of reading aloud the plays and sonnets at home. He attended actors as a physician and never charged them a fee. In appreciation of these services, the managers of the Covent Garden Theatre presented him with a handsome silver cup in 1849.

John Forbes continued to support his friend, John Conolly, in his work as Superintendent of Hanwell Asylum and, later, at Earlswood National Training Asylum for mentally defective children near Redhill. Its "able Medical Superintendent", Dr. John Langdon-Down (1828-1896), achieved eponymous fame by describing the condition then known as "Mongolism" and now called "Down's Syndrome". Forbes regularly attended meetings of the Board of Management until two years before his death in 1861. Conolly was famous for his pioneer treatment of mentally disturbed patients without any form of mechanical restraint. In this, he followed the teachings of the Quaker, William Tuke (1732-1822) at the York Retreat and Philippe Pinel (1745-1826) in France.

Forbes returned to Chichester in 1850 to lecture at the local

Literary and Philosophical Society, which he had founded in 1831. His subject was the theme for his MD thesis at Edinburgh (1817) *"Of Happiness in its relation to work and knowledge"*. His lecture was published privately in Chichester in 1850 and again in 1867. In 1852, he was conferred with an honorary degree of Doctor of Civil Law by the University of Oxford and, in 1853, he was knighted.

On his retirement from medical practice in 1848, Forbes took a well deserved walking holiday in Switzerland: he described this in a book entitled *'A Physician's Holiday or a Month in Switzerland in the summer of 1848'*. This was published in 1849 and included a map and illustrations of his prodigious journeys; in the Forbes' customary style, there are appropriate quotations from Roman and English writers and poets. He dedicated the work to his elder brother Alexander, whom he had not seen for more that 30 years (Forbes, senior, was then living in Mexico); to his credit, John had kept up a regular correspondence and he acknowledged the moral support he had received from Alexander throughout this time. The travel book, like the two others that followed in 1853 and 1855, contains much valuable information, even for modern readers, on the scenery and problems he encountered en route. Published by John Murray, it ran to a second edition in 1850 and a third in 1852.

Forbes had written to Sir Robert Peel (1788-1850) in 1845, the first year of the famine in Ireland, in support of cheap corn. In August 1852, he decided on a visit in order to see for himself the state of the country in the immediate post famine years. *'Memorandums made in Ireland in the autumn of 1852'* was published in two volumes and gives a detailed account of his journey throughout the whole island. Impartial accuracy is the essence of the book, in which he felt that he had "been animated by a sincere desire to speak the truth, and to do good according

to my humble means". This book has been documented else-
where and may be a further example of Forbes' outspokenness
being a source of dissension.

In 1853 the author Charlotte Brontë (1816-1855) acknowledged receipt of a box of books which included Forbes'
Memorandums and commented that she had read them with
"great pleasure". The publishers were Smith, Elder & Co., as
they were for his third travel book, *'Sight-seeing in Germany and
the Tyrol in the autumn of 1855'*. This was printed in two editions
in 1856. This travel work was probably his best in terms of his
descriptions of architecture and life in the countries visited by
him. His journey took place between the end of July and 25th
September, 1855. Forbes kept a journal in the meticulous
fashion inculcated during his time in the Royal Navy: he
recorded his impressions of such cities as Berlin, Prague, Vienna
and Paris. He commented very favourably on the art galleries
on the Continent compared to the National Gallery in London
and estimated his travelling costs at about twenty-four shillings
a day or less.

Charlotte Brontë had consulted Forbes about Anne's (1820-
1849) pulmonary tuberculosis but although he did not make
the long journey to Haworth to see Anne during her terminal
illness in 1849, Charlotte was so impressed by his humanity that
they exchanged copies of their books; in Charlotte's case she
inscribed a copy of *Villette* with the words 'in acknowledgement
of Kindness.'

Forbes was a prolific letter writer to his contemporaries: copies
of letters that he wrote to such celebrities as Sir Walter Scott
(1771-1832), Charles Dickens (1812-1870) and Florence
Nightingale (1820-1910) are among his papers.

Although apparently fit in the early autumn of 1855, Sir John had been unable to go out to Smyrna (Izmir) to set up a military hospital towards the end of 1854: he made some initial plans to do so in order to help stem the egregious mortality resulting from the campaign in the Crimea (1854-1856) but finally backed out for health reasons. One may speculate that he also hesitated to become involved in the squabbles regarding the selection of medical and surgical staff for Smyrna. He missed, albeit unknowingly, the opportunity of meeting Dr. Arthur Leared (1822-1879) - later to invent the bi-aural stethoscope - as the Irish physician had been appointed to the Civil Hospital there in 1855. Sir John Forbes did not work there, nor at the Civil Hospital at Renkioi, which was opened in October, 1855, so did not share duties with its medical superintendent, Dr. Edmund Parkes (1819-1876). Parkes had been editor of the *'British & Foreign Medico-Chirurgical Review'* from 1852 to 1855 and was later an outstanding and innovative army professor of military hygiene; he was author in the *'Review'* of the posthumous tribute to Sir John in 1862.

The British civil hospitals at Smyrna and Renkioi were staffed by female nurses independent of Florence Nightingale at Scutari, but, unlike his friend Sir James Clark, Sir John was never closely involved with her reforms in the living conditions of the British Army and nursing standards. They were neighbours in Old Burlington Street, where she lived at No. 30 (Burlington Hotel annexe) from November 1856 until Forbes retired to Whitchurch in 1859. In a letter dated 23 February 1860 (West Sussex Record Office, ADDMS No. 2545), she requested from Sir John a copy of his book *'Of Nature and Art in the Cure of Disease'* as, she remarked, his book reinforced her own views on the value of bedside observation. It is clear from this correspondence that her letter was in reply to one that she had recently received from Sir John about her 'Notes on Nursing' book but that illness had delayed her reply. (See full text of letter on pages 65-69)

Forbes' final publication, *'Of Nature and Art...'* (1857) was a philosophical little book based on his favourite theme of 'vis medicatrix naturae'. In twelve chapters, he sets out his case based upon his original ridiculed article (1846) in the *'Review'*. Sir John's book was well received, not only at home but also in the United States of America. There is a signed copy, presented by John Forbes, in the library of the London College of Physicians. It ran to a second English edition in 1858 and a Swedish translation was published in the same year. Although long-winded in Victorian fashion, the work could be read with advantage by modern doctors as well as by the general public.

In spite of Forbes' plea for a more rational approach to the control of disease, it is interesting to note that by the year of his death in 1861, there were still an estimated 50,000 patients in workhouse wards compared to 11,000 in voluntary hospitals. (Innes Williams, 1999).

Fig 6: Sir John Forbes in later life.

Chapter 8
Closing Years

In 1859, Forbes decided to retire from public life and to live at the home of his only son, Alexander Clark Forbes, at Whitchurch-on-Thames. The reason was that, for the previous two years, he had suffered from transient cerebral ischaemic attacks causing giddiness and falls. In a letter to Dr. William Munk, Treasurer of the Royal College of Physicians, from 'Whitchurch-Reading' and dated 4 May 1860, Forbes offered his resignation from the Comitia of the College on the grounds that: "I am a poor chair-ridden invalid, having totally lost the power of self-locomotion". (Archives of RCP Lond., Gen. Coll. A.L.S., quoted by kind permission of the Librarian). These symptoms were a sign of hardening of the arteries. He was also said to have suffered from a chronic breathing complaint and enlargement of the heart. Knowing that his health was failing Sir John, in March 1859, had donated his large library comprising between 3,500 and 4,000 volumes in English, French, German and Latin to his old Alma Mater, Marischal College at the University of Aberdeen. Sir John, with native shrewdness, stipulated that any expense incurred in removing the books should be defrayed by the University: this was agreed at a meeting of the Senate. (College Minutes, 5th April, 1859).

For the last two years of his life he lived in the tranquil rural surroundings of Swanston House, near Reading. Also in residence were his 36-year-old barrister son and his wife with their five children and eight servants. The household at Whitchurch also included his elder brother, Alexander, who had returned home from Tepic in Mexico to live with Sir John in Old Burlington Street some years previously. Across the road from Swanston House was the parish church: there is no record that Sir John ever attended church regularly like the rest of the family, as he probably preferred more practical forms of Christianity by helping deserving causes. Doubtless he would have been sympathetic to the sentiments expressed in a later eminent Scottish physician's prayer:-

'From inability to let well alone, from too much zeal for the new and contempt for what is old, from putting knowledge before wisdom, science before art and cleverness before common sense, ... good Lord deliver us.'

One account of Sir John's final illness states that he was completely paralysed for the last three months of his life but that he was visited regularly by his faithful life-long friend, Sir James Clark. His obituary notice in the *'Vienna Medical Wochenschrift'* refers to attacks of dizziness and loss of consciousness, which progressed to memory impairment and a tendency to fall to the right. He died peacefully from terminal infection in the evening of 13 November, 1861, just before his seventy-fourth birthday. The cause of death was certified as "progressive Softening of Brain and disease of its vessels - 5 years. Carbuncle-14 days".

Apart from the scathing obituary notice in the *'Lancet'*, others were very favourable to his memory. This was not surprising in the case of the *'Medical Times & Gazette'* as its late editor, F. Knight Hunt (1814-1854), had been a close friend of Dr. John Forbes, who had attended Hunt in his terminal illness in 1854.

A further appreciation of Forbes' life, including a description of his funeral at Whitchurch attended by his old friends, James Clark and John Conolly, appeared in the *'Medical Times & Gazette'* one week later. A particularly good account of Forbes' career was printed in the *'Proceedings of the Royal Society'* and a shorter version in *'Munk's Roll of the Fellows of the Royal College of Physicians of London'* for 1844. Among the foreign tributes was a detailed eulogy of his life and medical career from Vienna, where he was well-known.

Details of Sir John's will have been described, (obituaries). His estate was valued at £8,000. (This amounts to about £400,000 at today's values of the pound). The sole executor was Alexander Clark Forbes of Whitchurch, to whom he left all his property with the notable exception of a small cottage attached to his old school at Fordyce in Scotland. (West Sussex Record Office, ADDMS 2643). This cottage was occupied at that time by a "John Robertson". Forbes clearly stipulated that the leasehold was to remain with John Robertson during his lifetime but that afterwards, it should pass to the minister of the "... Parish of Fordyce and the Master of the Parish School..." in trust in order that any appropriate rent should be used for the purpose of suitable books as prizes for "... distribution to the best behaved boys and the best scholars at the said school...". The award of these prizes was left to the decision of the Parish Minister and the Schoolmaster who, if they disagreed, were to call upon "... any Minister or Clergyman or other important person residing in the said County of Banff..." to act as adjudicator.

In December 1861, Forbes' son arranged for his mother's remains to be removed from Kensal Green to be reinterred with those of her husband in a new family grave in St. Mary's churchyard, Whitchurch. He also placed memorial plaques to each of his parents side by side in the interior of the church. The text for Sir John's reads "Cast thy bread upon the waters: for thou shalt find it after many days. *Ecclesiastes X1 1*". A tribute to his late father's life by Alexander Clark Forbes, inscribed on the plaque, concludes: "... He passed through life without reproach, and died as he had lived, an honour to his noble profession".

Sir John was esteemed at home in many ways and abroad his merit was recognised by the award of the membership of a large number of renowned academic societies in both Europe and America. These included: Madrid, Göttingen, Vienna, Turin, Amsterdam and the Philosophical Society of America in Philadelphia; others were in Paris, Massachusetts, Brussels, Lisbon, Florence, Guadalajara (in Mexico), New York, Stockholm and Hamburg.

Fig 7: Plaque of Sir John Forbes in Whitchurch Parish Church.

Chapter 9
Epilogue

Readers of this booklet may be surprised that there is no mention of any incidents relating to Forbes' duties as Court Physician and especially to Prince Albert (1819-1861), who was given the title of Prince Consort by Queen Victoria in 1857. Royal protocol and medical etiquette may well have prevented any anecdotes but, almost certainly, the advice of his Court Physician would have influenced the Prince Consort's diplomatic promotion of social and industrial reforms.

Although as a physician he was not a great innovator, Forbes still deserves to be remembered as an outstanding medical journalist, a humane doctor and diligent scholar and, above all, as a translator of Laënnec in 1821. Considering that this work was done in a remote corner of Britain, with only his own personal knowledge of French to call upon, this was a remarkable achievement for a provincial physician. He built on this by his further translation of Auenbrugger and observations on the practical use of stethoscopy at Chichester at a time when the stethoscope was relatively unknown and unpopular. John Forbes lead the way in Britain by pioneering Laënnec's discovery just as Pierre-Adolphe Piorry (1794-1879) in France and Austin Flint (1812-1866) in America, later followed.

It was William Stokes (1804-1878) of Dublin who wrote the first systematic description of the use of the stethoscope while still a medical student at Edinburgh in 1825. In his preface Stokes acknowledged not only the prior "... translation of M. Laennec's work by Dr. Forbes, but also of a work on the application of the stethoscope, by the same distinguished physician;". From this it would appear that the Dublin author was well aware of the previous Scottish contribution to the art of stethoscopy.

At the age of 52, Forbes had embarked on a risky career in medical journalism in London which, although controversial at times, contributed in large measure to the promotion of sound medical literature in the nineteenth century. He lived to see his opinions respected and even the 'Lancet' posthumously admitted that he was a "benevolent and conscientious physician".

His career as Physician to the Royal Household coincided with that of Sir James Clark. The wise advice of both was available to Prince Albert, who planned the Great Exhibition of 1851 at a time of great industrial expansion and advances in public health. Nursing reforms were the prior concern of Florence Nightingale. It is clear from her correspondence with Forbes that they respected one another's opinions on matters of the bedside observation of sick patients.

His philosophy of life may be epitomized by the motto he adopted for his coat of arms: LABORE ROBORE SPE. *(By Work by Strength by Hope)* Perhaps this booklet will help to ensure that Sir John's achievements will survive into the twenty-first century.

List of John Forbes' ships in which he served in the Royal Navy from 1807 to 1815.

References give details of type of ship, armament etc. and further information from Public Record Office.

1) HMS ROYAL GEORGE. (Ref: Lyon D.) [see below]; Part 1, Chapter 4, The Slade Era, 1745-1785, page 63).
2) HMS ROYAL WILLIAM. Part 1, Chapter 3, page 39.
3) HMS CASTOR. Part 1, Chapter 4, page 84.
4) HMS ABERCROMBIE. Part 2, Chapter 7, page 270.
n.b. ABERCROMBIE was the French Prize D'HAUTPOULT - see J Med Biog 6; 63-67, number 2: May 1998.
5) HMS NETLEY. Part 2, Chapter 7, Prizes & Purchases, page 279.
6) HMS CHERUB. Part 1, Chapter 5, page 130.
7) HMS VIMIERA. Part 2, Chapter 7, page 279.
8) HMS DÉSIRÉE. Part 2, Chapter 6, page 224.
9) HMS BENBOW. Part 1, Chapter 5, page 114.
10) HMS VENERABLE. Part 1, Chapter 5, page 113.

References:
I) Lyon D. *The Sailing Navy List: All the Ships of the Royal Navy - Built, Purchased and Captured, 1688-1855.* London: Conway Maritime Press, 1993.
II) John Forbes (2); *Public Record Office, ADM 104/30, 509.*

From April 2003 the official accreditation of the Public Record Office (PRO) became the National Archives (NA).

Letter from Florence Nightingale to Sir John Forbes.

30 Old Burlington St,
London W

Feb 23/60

My dear Sir

Nothing has given me half so much pleasure as a note from you about my little Nursing book. That you, to whom the world is so much indebted in the matter of its health, should endorse it with your imprimatur is a very great satisfaction to me.

All I can say for the book is that there is not one word of theory in it. Every sentence of it is the fruit of bitter experience. That your experience as a Physician should so coincide with mine as a nurse gives it value.

The great object I had in view was to recall the art of observation which has, I think, deteriorated, even in my day, under the load of supposed science. People have eyes "and they see not".

My conclusions were arrived at by looking at disease simply from the practical side. If people who have Science too, (which I wish I had,) would do the same, how much might not be done for the World's

I know your book "Nature & Art in the cure of Disease" well. But should it not be a trouble to you to send me a copy, as you kindly offer, I should consider it a great honor [sic] to have one from you.

I should have answered your kind note before, had it not been for illness. Believe me I remain,

dear Sir John
faithfully and gratefully yours

Florence Nightingale

PTO

Letter from Florence Nightingale to Sir John Forbes

She added a postscript: 'You encourage me by your kindness to send you another little book of my Hospital experience.' [Author's note: It seems very likely that "my little Nursing book" was FN's classic *'Notes on Nursing'*, published in December 1859. This became very popular and it is not surprising that a copy had reached Sir John at Whitchurch.

The second "little book" probably refers to FN's *'Notes on Hospitals'*, which was also published earlier in 1859 by J.W. Parker & Son, London: this work drew attention to the unhygienic and overcrowded conditions on the wards, which gave rise to higher mortality in hospitals than in equivalent populations in the community. In 1862, Florence herself commented on the state of a Liverpool workhouse that it was "worse than Scutari!" Miss Nightingale's advice was much sought-after in the designing of new hospitals throughout Great Britain and abroad as a result of her book, which ran to further editions in 1860 and 1863].

Select Bibliography

R. A. L. Agnew, A.J. Larner, Dr John Forbes and the Brontës, *Medical Historian; 27 (2017)*, 37-42.

Agnew RAL: *Forbes in Hibernia: the narrative of a Scottish Physician's visit to Ireland in 1852*. J Ir Coll Phys Surg, 21 (1992), 40-44.

Agnew RAL: *John Forbes (1787-1861), in memoriam: from Cuttlebrae to Whitchurch*. J Med Biog, 2 (1994), 187-192.

Agnew RAL: *John Forbes and Daniel O'Connell: the meeting of a Scottish Physician with an Irish nationalist in 1841*. J Med Biog, 4 (1996), 178-183.

Agnew RAL: *Fortune favours the brave - the Capture of Guadeloupe 1815*. J R Nav Med Serv, 83 (1997), 94-98.

Agnew: RAL: *"All that glisters is not gold". Sir John Forbes (1787-1861): a West Indian enigma*. J Med Biog, 6 (1998), 63-67.

Agnew RAL: *Sir John Forbes (1787-1861) and Miss Florence Nightingale (1820-1910)*. Vesalius VII (2001), 36-44.

Agnew RAL: *Catalogue of the Library of Sir John Forbes (1787-1861)*. J Med Biog, 9 (2001), 104-108 & 175-180.

Agnew RAL: *The Capture of Guadeloupe in 1815 - The Role of John Forbes (1787-1861)*. The Nelson Dispatch, 12 (2017), 681-695.

Anon. Lancet, 5 No. 5, (1824), 144-160; 169-180.

Anon. Edinburgh Medical & Surgical Journal, 23, (1825), 406-416.

Bishop PJ: *The Life and Writings of Sir John Forbes (1787-1861)*. Tubercle, 42 (1961), 255-261.
[Includes list of four known portraits of Forbes]
Bishop PJ: *Evolution of the Stethoscope*. J R Soc Med, 73 (1980), 448-456.

Blair JSG: *The New College and professional education*. In: The Centenary History of the Royal Army Medical Corps (1898-1998), Edinburgh, Scottish Academic Press (1998); 4, 95.

Brody J: *An émigré physician: Dr. David (Didier) Roth, homeopath, art collector, and inventor of calculating machines*. J Med Biog, 8 (2000), 215-219. [Includes good overview of homeopathy]

Bynum WF and Wilson Janice C: *Periodical Knowledge: medical journals and their editors in nineteenth-century Britain*. In: Medical Journals and Medical Knowledge, Historical Essays, ed. by WF Bynum, Stephen Lock and Roy Porter. London and New York: Routledge, (1992); 2, 41-43.

Clark J: *Medical Notes on Climate, Diseases, Hospitals, and Medical Schools, in France, Italy and Switzerland, comprising an Inquiry into the Effects of a Residence in the South of Europe in cases of Pulmonary Consumption and illustrating the Present State of Medicine in those countries*. London, T & G Underwood (1820).

Clark J: *A Memoir of John Conolly, M.D., D.C.L.*, London, John Murray (1869).

Cormack AA: *Two Royal Physicians: Sir James Clark, Bart., 1788-1870, Sir John Forbes, 1787-1861: Schoolmates at Fordyce Academy*. Reprint from the Banffshire Journal (26 June 1965).

Coakley D: *Arthur Leared (1822-1879)* - Inventor of the bi-aural stethoscope. In: Irish Masters of Medicine, Dublin, Town House, (1992); 23, 189-191.

Craig J: *A general dispensary practice 150 years ago*. Reprint from the Aberdeen University Press Review (Autumn 1972), 44, No. 148, 358-367.
[Includes a description of Forbes in Penzance]

Forbes J: *A Treatise on the Diseases of the Chest in which they are described according to their Anatomical Characters, and their Diagnosis established on a new principle by means of Acoustic Instruments.* London, T & G Underwood (1821).

Forbes J: *Original Cases with Dissections and Observations illustrating the use of the Stethoscope and Percussion in the Diagnosis of Diseases of the Chest; also commentaries on the same subjects selected and translated from AVENBRUGGER, CORVISART, LAENNEC and others.* London, T & G Underwood (1824).

Forbes J: *Of Nature and Art in the cure of Disease.* London, John Churchill (1857. Second edition 1858).

Foster RF: *Land Politics and Nationalism.* In: Modern Ireland 1600-1972, London, Allen Lane, The Penguin Press (1988) 16, 396.

Goldie Sue M: *Florence Nightingale, Letters from the Crimea 1854-1856.* Mandolin: Manchester Univ. Press (1997).

Greenhill WA: *Forbes, Sir John (1787-1861).* In: Dictionary of National Biography, ed by L. Stephen and S. Lee. London: Oxford University Press for Spottiswoode & Co., (1967-1968), 7, 405-407.

Grosskurth P. *The Byrons - Impetuous, Bad and Mad.* In: The Flawed Angel, London, Hodder and Stoughton (1997), p 17.

Hunt JL: *Untamed editor: F Knight Hunt MRCS (1814-1854).* J Med Biog, 5 (1997), 210-220.

James W: *Colonial Expeditions - West Indies [1810].* In: The Naval History of Great Britain, from the Declaration of War by France in 1793 to the Accession of George IV, London, R Bentley (1859).

Johnson James: Obituary in Munk's Roll of the Royal College of Physicians, 2nd Edition, Revised and Enlarged. Vol. III, 1801-1825. Published by the College, London, (1878), pp 238-241.

Laënnec RTH: *'De L'Auscultation Médiate; ou, Traité du diagnostic des maladies des poumons et du coeur, fondé principalement sur ce moyen d'exploration'*. 2 vols., Paris: Brosson et Chaudé, (1819).

Longford E; *Victoria RI* Illustrated Edition, chapter 8, *Mama's Amiable Lady 1839*. London Weidenfeld and Nicolson, 1973.

Lyons JB: *The Irish Journal of Medical Science - a historical outline.* Ir J Med Sci, 169 (2000), 143-148.

Major RH: *Classic Descriptions of Disease.* Springfield (Illinois), Baltimore (Maryland), Charles C Thomas (1932).

Parkes EA: *Memoir of Sir John Forbes Kt*...Reprinted by permission, from the January Number, 1862, of the British & Foreign Medico-Chirurgical Review (For Private Circulation)'. By EA Parkes, with preface by Alexander C. Forbes, 1 (1862), 7-70. [Includes anecdotes of Forbes' naval career].

Sakula A: *Sir John Forbes (1787-1861). A bicentenary review.* J R Coll Physicians Lond, 21(1987), 77-81. [With complete bibliography].

Sakula A: *Laennec's influence on some British physicians in the nineteenth century.* J R Soc Med, 74 (1981), 759-767.

Shepherd J: *The Civil Hospitals in the Crimea (1855-1856).* Proc R Soc Med, 59 (1966), 199-204.

Steer FW: *The Royal West Sussex Hospital..The First Hundred Years 1784-1884.* The Chichester Papers, Chichester City Council, 15, (1960).

Stokes W: *An Introduction to the use of the Stethoscope with its Application to the Diagnosis in Diseases of the Thoracic Viscera; including the Pathology of these Various Affections.* Edinburgh, MacLachlan and Stewart (1825).

Strawson J: *The Duke and the Emperor - Wellington and Napoleon.* London, Constable (1994).

Taylor S: *John Keats (1795-1821).* J Med Biog, 2 (1994), 209-216.

West Sussex Record Office: ADDMS 2543. Gives date of John Forbes' appointment to the Royal Household.

Tuke, William (1732-1822). In: *Chambers Biographical*

Dictionary. Gen. Ed M Magnusson, KBE, Asst. Ed. R Goring. Fifth Edition (1990), Edinburgh: W & R Chambers Ltd., p1478.

Woodham Smith C: *Florence Nightingale, 1820-1910.* London, Constable (1950).

Williams Harley: *The Healing Touch.* London, Jonathan Cape (1949), p 75.

Williams Innes D: *The Poor Law Infirmaries.* In: 27th Henry Cohen Lecture on the History of Medicine, delivered at the Liverpool Medical Institution on 3rd November 1999. Liv. Med.Inst. Transactions and Report (1999-2000); 9.

W.W.W. : *Parkes, Edmund Alexander (1819-1876).* In: Dictionary of National Biography, L. Stephen & S Lee eds., London, Oxford University Press for Spottiswoode & Co., (1967-1968), 15, 294-296.

Additional SELECT BIBLIOGRAPHY:

Coulings S: *History of the temperance movement in Great Britain and Ireland.* London (1861), William Tweedie, P296.

ALSO: Egan J: John O'Neill. Recorded as a Shoemaker and Temperance Poet, but More to be Valued, perhaps, as a Valuable Witness to the struggles of the Irish Poor in London. *The Irish Genealogist,* (2005), Vol.II, p 288.

Egan J: A STRANGER WHO UNDERSTOOD from *Memorandums made in Ireland in the Autumn of 1852 by John Forbes.* Ibid pp 300-02

Barker J. *The Brontës.* London (1994), Weidenfeld & Nicolson, pp 690, 843.

Lloyd C & Coulter JLS. *Medicine and the Navy 1200-1900. Edinburgh & London 1961: E&S Livingstone Ltd, 1V 202-3.*

Smith M.(ed.), *The Letters of Charlotte Brontë* .Vol. III p.170.

Smith WDA: *Franz Mesmer (1734-1815) and Mesmerism.* In: Henry Hill Hickman. Sheffield (2005); JW Northend Ltd: 3 p 25. With acknowledgement to Dr Adrian Padfield, Sheffield.

Virgil's Aeneid: 10, 1, 284. In: *The Oxford Dictionary of Quotations.* Revised Fourth Edition. Ed. Angela Partington 1996, Oxford and New York: Oxford University Press: p 714, quotation 7.

Wikipedia – Royal Literary Fund: *A Short History,* by Janet Adam Smith, President RLF 1976-1984. HYPERLINK "http://www.rlf.org.uk" http://www.rlf.org.uk Accessed 16/06/08.

Obituaries of Sir John Forbes.

British Medical Journal, 2, 561-562 on Nov. 23, 1861; also a summary of his will on p 675 of issue of Dec. 21, 1861, Taken from the 'Illustrated London News'. See also: West Sussex Record Office ADDMS 2543.

The Lancet, 2, 512 on Nov. 23, 1861.

Medical Times & Gazette, Nov. 16, 1861, 504-507; also an account of his funeral held on Tuesday, 19 November at Whitchurch in the issue of Nov. 23, 1861. Vol. IV (1826-1928) Proceedings of The Royal Society of London 1862-1863, Vol 12, pp 6-11. It also mentions his generous bequests of £100 each to the Medical Benevolent Fund and College. p.535.

'Lives of the Fellows of the Royal College of Physicians of London 1826-1925', compiled by GH Brown, published by the College in 1955 and commonly known as 'Munk's Roll': entry for the year of Forbes' Fellowship [1844], 'Vol.IV' Fellows 1826-1925, pp 34-35.

Anonymous. Nekrolog. Sir John Forbes, geb. 1787, gest. 13 Nov. 1861. *Wien Med Wochenschr 1861*; 11: 789-90. This gives an excellent account in German of Forbes' life and contributions to medical knowledge.